GREENACRE, D. Magic
Lanterns
MB 8/86   1.56

8412

HERTFORDSHIRE LIBRARY SERVICE

Please return this book on or before the

last date shown or ask for it to be renewed

MAGIC LANTERNS

Derek Greenacre

Shire Publications Ltd

CONTENTS

Set in 9 point Times roman and printed in Great Britain by C. I. Thomas & Sons (Haverfordwest) Ltd, Press Buildings, Merlins Bridge, Haverfordwest.
British Library Cataloguing in Publication data available.

ACKNOWLEDGEMENTS
    The author acknowledges with gratitude the assistance of the following in the preparation of this book: I. Greenacre, G. O'Connor, P. Banham, M. Smith and F. Branson. Photographs on the following pages are acknowledged to: Keith Brown, pages 1, 3, 4, 11, 20, 21, 22 (lower), 23, 24, 25 (upper), 27 (upper), 28, 32; George Skipper, cover and pages 2, 6, 7, 8, 9, 10, 12, 14, 16, 17, 18, 19 (lower), 22 (upper), 25 (lower), 27 (lower), 29, 31. All other illustrations, including the line drawing on page 5, are by the author.

COVER: *A mahogany magic lantern by the London firm of Newton and Company. The chimney of the lantern is unusual, being made of polished copper rather than tinplate.*

BELOW: *A magic lantern by an unknown maker. It had never been used and had remained in its box since about 1900. Consequently it is in pristine condition and gives a good indication of what a new lantern would have looked like in Victorian times.*

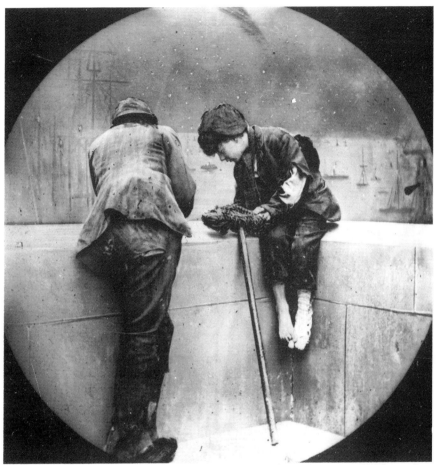

A scene from 'Road to Heaven' by G. R. Sims. The boy is about to jump into the Thames in an attempt to escape his earthly plight and reach heaven. He awakes in hospital and mistakes the white-robed doctors for angels.

A THRILL IN THE DARK

Most people have had the experience of sitting in a darkened room and seeing how exaggerated shapes are formed on the walls by the flickering firelight. The experience is strange in that it fuses two opposite emotions. The feelings of warmth and security are tinged with a hint of the mysterious and the supernatural; it is a world of half real and half imagined images. This experience must go back to the days of the caveman and it is not difficult to imagine these images

being incorporated into stories. The stories and the shadows would eventually be developed into ritual, magic and superstition and these in their turn would demand an even greater need for refinement if the ceremony was to be associated with the spirit world. With even simple props, shadow images can be modified and elaborated into complex art forms and, even today, there are entertainers who still make a living merely by manipulating their hands to form shadow

LEFT: *A galanty shadow show. This type of show explored the potential of the projected image. This illustration comes from the 'Boy's Own Paper' of 1889.*
RIGHT: *A magic lantern slide of a Punch and Judy show. Many early slides drew on earlier forms of entertainment for ideas but Punch and Judy was probably the first type of entertainment to be performed by shadow shows in Britain.*

pictures on a screen.

Travelling puppet shows (notably the well documented Punch and Judy) toured Britain and the fact that the show is still flourishing is proof of its popularity. On darker evenings, however, the show presented problems and it was not long before some enterprising showman had the idea of stretching a thin cloth over the front of the puppet theatre, placing an oil lamp behind it and arranging the puppets between the two. When viewed from the front the puppets then appeared as silhouettes on the screen and complicated effects could be achieved by movement of puppets relative to the light source. These shadow shows of Britain had their eastern counterparts and, as early as the eleventh century, complex shadow shows were to be found in China. The puppets appearing in such shows were usually two-dimensional, being constructed of animal skins scraped thin and made translucent by immersion in oil.

The shadow theatre evolved into other areas but continued to co-exist with its later counterparts. The light used in shadow shows became more efficient, firstly by means of reflectors and later

with the addition of magnifying lenses. The popularity of the shadow shows reached its peak in Europe in the latter years of the sixteenth century but even in the late nineteenth century the concept still generated enough interest to feature as a regular type of article in boys' magazines of the period. The shadow show had by then assumed the popular name of 'galanty show', a corruption of 'galante show', meaning that it was free from coarser elements.

With the increasing use of lenses and light sources, it was inevitable that a combination of light source, image and lens which would improve the format of the picture would be discovered. One of the first people to make use of this development was a Jesuit priest called Kircher, who, as early as the mid seventeenth century, referred to a device for projecting coloured images on to a screen by means of a lens. Whether Kircher invented the device is uncertain but, without doubt, the apparatus took the image away from being a silhouette and all subsequent devises were merely refinements of this basic format.

The three essential elements of the

magic lantern had evolved into: a light source; a transparent image painted on glass; a lens to focus the image into a sharp realistic picture on a screen. When these three components had been assembled in the correct order the magic lantern took on its true form and only slight technical modifications made it differ from a modern slide projector.

Since the slide projector had been invented as early as the mid-seventeenth century it seems remarkable that such a versatile piece of equipment should be scarcely made use of until the golden age of the magic lantern in the latter part of Queen Victoria's reign.

The failure of the early magic lanterns to achieve universal popularity was for three main reasons.

Firstly, the early projectionists kept the existence of the lantern secret as they deliberately used the device to 'conjure up spirits'; the fewer people who learnt of the hidden apparatus, the more grotesque images could be attributed to the sorcerer's magic powers. The vast majority of early magic lantern slides (from which we may infer the vast majority of magic lantern shows) depicted demons

ABOVE: *Shadow show character of 'The Ghost'. Such tongue-in-cheek links with the supernatural were common in many early forms of folk entertainment.*
BELOW: *Parts of a magic lantern: (a) objective lenses; (b) focus knob; (c) filter slot; (d) flasher; (e) condenser lenses; (f) chimney; (g) door; (h) lime; (i) gas pipes; (j) slot for slides; (k) base board.*

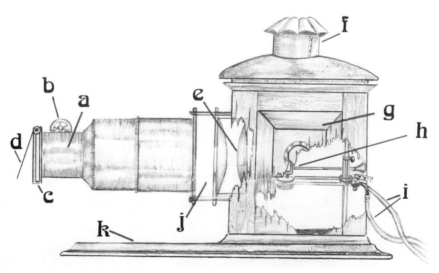

and spirits. Indeed, when a large exhibition of magic lantern material was to be assembled in London in 1964, it allowed the organiser of the exhibition to view the whole history of the magic lantern and, in the catalogue from the exhibition, entitled 'Origins of the Cinema', he described the earliest as 'notably sadistic in subject'.

The second reason for the failure of the early magic lantern to achieve universal success was that, before the invention of photography, slides had to be laboriously hand-drawn and hand-painted. Although this would appear to be an ideal cottage industry, the great difficulty in painting the slides prevented this from developing. Because the magic lantern magnified the image so enormously, any slight discrepancy in either outline or thickness of paint showed up as a serious defect when projected on the screen. To compensate for this difficulty many early slides were drawn on larger sheets of glass and, in order to prevent breakage, these were often bound in wooden frames. These frames had the dual purpose of making the glass safer to handle and preventing the image from being scraped off the slide as it was placed in the lantern. Unfortunately, the wooden frame added to the cost and, even with the introduction of jigs to cut and assemble joints accurately, mass production was still difficult.

The third problem was that of the light source. The early magic lanterns were lit by either a candle or a small oil lamp, and neither was able to produce a bright enough picture more than a few feet away from the screen. This technical limitation made viewing by a large audience impossible.

By the start of the Victorian era the first two problems had been eliminated. Firstly, the lantern's powers could no longer be attributed to 'magic' (although the name stuck). Secondly, the photographic process had been evolved, largely through the work of Fox Talbot. This not only enabled photographs to be produced on glass but also, because a negative was used, allowed duplication of the same image.

A solution to the third problem of producing an efficient light source proved much more elusive; there were many false starts and many blind alleys.

An oxygen/hydrogen lime burner for the magic lantern. The complicated system of control knobs allowed the lime to be moved in any one of three dimensions.

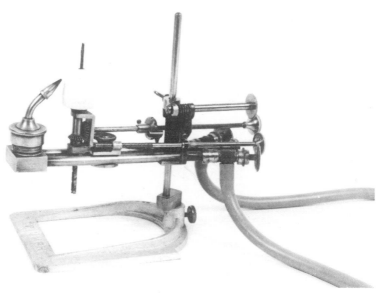

The Newtonian variation of the Sciopticon paraffin burner. The photograph shows the burner in the open position to reveal the four linear wicks and control knobs. Once lit, the burner would be closed, placed in the lantern and the chimney added.

ILLUMINANTS

The search for efficient illumination was not restricted to magic lantern applications. Indeed, most research had already been pioneered in the quest for suitable lighting in the home.

Both the candle and the oil lamp were well known in antiquity but neither produced a flame which burnt as a truly effective light source. By adding a chimney to the oil lamp, an updraft of air allowed increased oxygen to produce a brighter flame. It also had a second advantage in that it reduced intermittent draughts, which produced a flickering effect. In the case of the magic lantern illuminant, the chimney was to prove essential for a steady image.

However, the lack of power of the oil lamp, together with the amount of heat generated by it, proved irksome and prevented mass viewing; glass slides frequently broke through thermal shock. The problem was reduced somewhat by the invention in America of a more efficient burner known as the Sciopticon.

This paraffin burner for the magic lantern comprised two linear wicks placed side by side and at right angles to the condenser lenses.

The basic idea of this burner was modified in England by the Newton lantern company, which added a third (and later a fourth) wick. More importantly, this burner was an encased unit which was removable from the lantern. This system had two advantages: it allowed lighting systems in magic lanterns to be interchangeable, and the heat was either contained or dispersed through the lighting system, rather than through the body of the lantern.

The design of the Newton-styled burner was well thought out. The chimney was made of metal rather than glass, thus preventing light escaping into the darkened room. The wicks themselves were encased at either end of the burner by removable squares of glass, which could easily be replaced should the glass crack. Being encased in the burner,

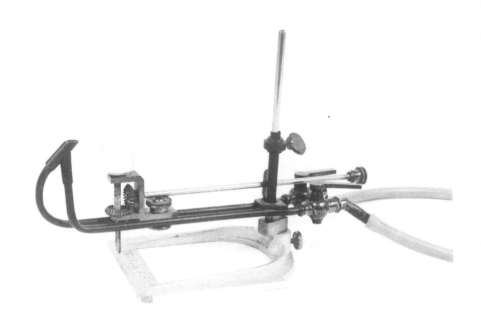

A blow-through jet which used coal gas and oxygen to ignite the lime.

the wicks had their air supply drawn up through them from below, thus greatly increasing the burning efficiency and consequently the quality of illumination. This, in conjunction with a parabolic reflector mirror, made the Newtonian burner the standard paraffin burner in almost all later lanterns.

Despite its increased efficiency compared with other burners, the Newtonian was less than perfect and by modern standards the picture was poor. The image on the screen had an orange-tinged glow, which gives it a nostalgic charm today. Badly handled, the paraffin burner produced a smoky flame, which could spread soot throughout the room.

In the early nineteenth century it was discovered that certain substances, notably lime, if heated to a sufficiently high temperature burn to produce a brilliant white light. The application of this principle to the magic lantern was not immediate. Indeed one of the most important uses of the discovery was as a means of signalling. The light produced by the burning lime was so powerful that it could be used to signal from Ireland across the sea to the west coast of Scotland. A better known application of the discovery was in the theatre, where burning lime was used for spotlights. This was the origin of the phrase 'in the limelight'. A magic lantern fitted with a lime burner was also often used in the theatre for special effects, notably the illusion known as 'Pepper's ghost'.

Despite its brilliance, limelight had the great disadvantage that the lime had to be heated to an extremely high temperature. Many methods were used to produce the heat, all of which involved some danger. Basically, the high temperature was achieved by a mixture of a combustible gas with oxygen: by this method the gas burnt efficiently and so increased the temperature. The commonest combination of gases employed to ignite the lime

8

ABOVE: *An early Victorian print of a magic lantern using a limelight burner. It is here being used to create the stage illusion known as 'Pepper's ghost'. Note the bent chimney on the lantern to prevent stray light from escaping.*
BELOW: *A selection of limelight equipment, including regulator valves, cylinder keys, ether cylinders, limes and lime tongs.*

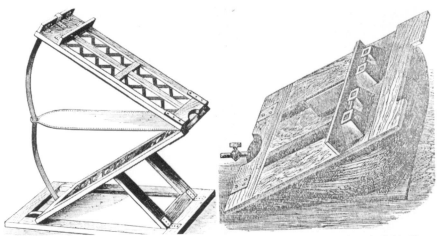

The gas used in limelight burners was contained in bags between pressure boards (left). Heavy weights were placed on the boards (right) to force the gas along the tubes connecting with the burner.

was a mixture of hydrogen and oxygen. In the early days rubberised bags were filled with gas and sandwiched between weighted wooden boards to create pressure but the bags were subject to natural wear and tear, making this a hazardous procedure.

The danger was reduced by the introduction of pressurised steel cylinders to contain the gases. These had to be fitted with regulators to deliver the gas at a steady pressure. The cylinders were both heavy and cumbersome and special keys had to be carried by the lanternist to operate the valves on them.

Cylinders too had their dangers, especially when handled without care. One incident occurred at a Bradford railway station when a local magic lantern company, Riley Brothers, sent a young boy in their employment to deliver some full cylinders. When taking two cylinders to the platform, the boy carried one and dragged the other. There was an explosion and he was killed.

The lime could be ignited by other means if coal gas was available in the house. Then, a specially designed limelight burner called the 'blow-through jet' could be used in the lantern but this too required an oxygen cylinder to increase the temperature of the burning gas.

Still more dangerous was the combination of ether and oxygen. These were combined in the ether saturator, but this device, even in the hands of experts, could be disastrous. The carnage caused by exploding ether at a magic lantern show in France resulted in both the government and the general public viewing the use of ether with disapproval.

The use of lime as an illuminant was altogether dogged with problems, the root of which was that the limelight was so brilliant that simpler methods of igniting it were sought rather than accepting that it was the high temperature of the ignition which caused the problem and not the ignition source.

The limes themselves were easy to transport, provided they were kept dry. As lime absorbs moisture from the atmosphere, the blocks of lime had to be kept in airtight containers. The best-quality limes can still be found today in workable condition because many were individually sealed in their own airtight glass tube. The lantern operator would usually take to the lantern show a pair of pliers, which enabled him safely to break the glass tube containing the lime block. The lime would then be placed on the burner and the tap of the cylinder of combustible gas turned on. The match would ignite the gas and then oxygen would be introduced to the flame. This increased the temperature of the burning gas until the lime flared into brilliance. It had then to be manoeuvred into the optimum position in the lantern body so that no shadow fell on the screen. The lanternist would be kept busy during a show for not only had he to place the slides in the lantern, but he had also to ensure that there was an ample supply of gas reaching the lime and that the lime itself was not all burnt away. Victorian lanternists used to express their contempt for over-talkative lecturers with the saying 'a gassy operator often runs short of it'.

Towards the end of the nineteenth century electricity was being used more often as a power source. The first attempts to illuminate the magic lantern electrically involved the use of an electric arc. Later, as specialist electric filament bulbs were developed, they were to replace other forms of lantern illuminants. Initially, these bulbs were used only because they were more convenient but eventually they were to rival the brilliance of the limelight.

A lantern slide of a banner advertising a 'Sacred Pictorial Concert' at the Central Hall, Newcastle upon Tyne. This building was used for many years to give magic lantern shows.

ABOVE: *A selection of cheaper lanterns made of black tinplate.*

BELOW: *A magic lantern with a 'Russian iron' body and brass lens tubes. The lantern was made by Perken, Son and Company, which used the trade name Optimus for its products.*

A home-made lantern by an unknown maker. The lantern has been constructed from materials which came to hand. The chimney, for example, has been made from an old chocolate tin.

MAGIC LANTERN TYPES

The earliest magic lanterns, which had been used to create the deception of 'conjuring up evil spirits', were often mounted on wheels and placed behind a translucent screen. As the lantern was moved further from the screen, the image grew larger. To the audience on the other side of the screen, it appeared as if the spirit was moving towards them.

As lanterns evolved into more acceptable and better known instruments, they became both more numerous and more varied. Their increased efficiency also boosted their popularity.

The cheapest lantern in Victorian times was a child's toy lantern. These could be bought for as little as a shilling but, compared to a real magic lantern, the picture and the quality of the slides available to fit them were poor. The commonest of these lanterns was made by the German firm of Ernst Plank and can be recognised by the trademark of a winged wheel, often with the initials *EP* below it. British firms, too, produced scaled-down lanterns for the young enthusiast but if a family could afford a full-size lantern it was preferable as slides for children could be purchased to fit the standard-sized lantern.

The cheapest of the full-sized lanterns was the home-made version, for which the manufacturers supplied the lenses but the body and construction were left to the purchaser to complete himself. Another inexpensive type was the black japanned tinplate lantern suitable for home entertainment. Lanterns of this style were very numerous in Victorian times, selling for about one guinea. Their popularity is reflected in the frequency with which they turn up in antique shops today (usually, however, without their chimney and paraffin burner, having been converted to electricity during the first half of the twentieth century).

13

ABOVE: *A small mahogany lantern made in Newcastle upon Tyne by Mawson Swan and Morgan. The convoluted chimney is typical of many lime-burning lanterns.*

BELOW: *The Phoenix lantern made by Reynolds and Branson of Leeds. This lantern is a fine example of Victorian craftsmanship, and the lantern body is made from polished walnut rather than the traditional mahogany.*

2704 The " **Phœnix** " optical lantern with walnut body, panelled side doors, complete in box £4 4 0

The Phoenix lantern as portrayed in the Reynolds and Branson catalogue of 1894.

The lantern of brass and 'Russian iron' was of slightly better quality. This was usually fitted with a limelight burner, although the lighting unit was interchangeable with that of a paraffin burner. The 'Russian iron' (which was actually made in Birmingham) was intended to be rust-resistant, a great advantage when using the jet of oxygen and hydrogen to heat the lime, for a by-product of this combustion was water in the form of steam. There were many variations in the body style of these lanterns but the shape was basically an aesthetic rather than a technical modification.

The best type of magic lantern was the metal-lined, wooden-bodied model. These elaborate devices were beautifully made in polished hardwood (usually mahogany) with superbly engineered fitted brass-bound lens tubes. The chimneys also took on a very ornate form, not merely as a manifestation of grandeur,

but principally because the large surface area of the convoluted metal helped disperse the intense heat of the limelight burners. Such lanterns were to be found in the more affluent homes and were traditionally associated with Christmas entertainment. Expensive lanterns would also be used by the professional lanternist, who was either himself an entertainer or a lantern operator for a public lecturer who wished to use the lantern to illustrate his talk. All professional lanternists had to cope with similar problems of projection and therefore tended to use similar equipment.

One of the main difficulties facing the professional lanternist would be coping with halls of different sizes. The magic lantern had to stand at the back of the hall to avoid blocking the screen, but as the lantern was moved further from the screen the picture became larger. In order to prevent the picture from spread-

A science lantern by Reynolds and Branson. As with ordinary lanterns, slides could be projected but the open area (seen in the side view, above) also allowed scientific experiments to be set up and projected on the screen.

A mahogany lantern on a Victorian projecting table. The table height could be adjusted to enable the picture from the lantern to fill the screen.

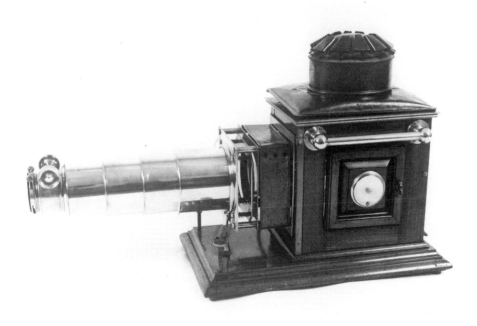

A large lantern by Reynolds and Branson of Leeds. This lantern was originally used by the firm to give lectures. Today it is used by the author to present magic lantern shows to the public.

ing over the edge of the screen, the lanternist had to install a different objective lens unit into the front of the lantern. The larger the focal length of the objective lens, the smaller the size of the picture produced. Lanterns with an interchangeable lens unit were therefore developed for professional use. Lenses of varying focal lengths had to be carried by the lanternist so that a hall of any size could receive a picture to fill the screen. A cheaper way of achieving a varying picture size was to buy a 'multi-focal objective lens' but these were usually avoided by the professionals.

As magic lanterns became more efficient, their potential as an educational tool was realised and it was not long before specialist lanterns were being produced by lantern firms. Many of these lanterns were also used by scientific lecturers who wished not only to present slides illustrating their talks, but also to enable the audience to witness scientific experiments. Special science lanterns were produced to permit experiments to be projected. These lanterns had an open area between the condenser lens and the focusing lens; this open area permitted the experiments to be assembled. The experiments were limited to transparent material, as opaque objects could be seen on the screen only in silhouette. Nevertheless, there was still much to display and such experiments as the decomposition of water by electrolysis were firm favourites. Very small objects could be displayed by means of a microscope attachment. However, tiny living organisms had to be treated with respect as the heat from the lime burner magnified through the condenser lens could easily kill them before their image had been projected. Special heat-absorbing tanks had to be introduced to prevent this.

2720

2720 Reynolds & Branson's Lantern Microscope, for projection of microscope slides on to the lantern screen, with sliding adjustment, differential screw fine adjustment, and rotating diaphragms £3 7 6

ABOVE: *The projecting microscope was responsible for many innovations in education, particularly biology. With this apparatus, the causes of diseases might be identified and brought to the attention of a mass audience.*

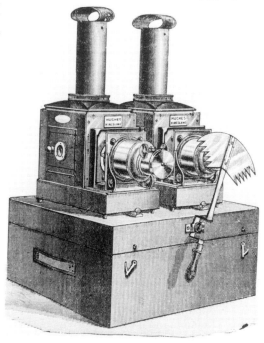

RIGHT: *A matching pair of magic lanterns used to create dissolving views by means of the curious fan-shaped device on the front of the stand. The biunial lantern was later to prove a more satisfactory means of achieving these effects.*

OPPOSITE AND ABOVE: *A set of slides entitled 'The Tiger and the Tub'. The bizarre ending is typical of many lantern stories. The slides were made by the London firm of Butcher and Son and sold under the trade name of Primus.*

MAGIC LANTERN SLIDES

As magic lanterns became more efficient and more diverse, so lantern slides evolved and grew more sophisticated. However, in early days the lens quality of the lantern did not match the quality of the slides, and later, as lens calculation and grinding became a more exact science, the lantern had to wait for the slide quality to achieve a similar perfection.

Early slides were usually large and bound in mahogany frames. The commonest of these was the panoramic strip, which was often crudely painted and came in long wooden boxes with sliding lids. It was, however, soon realised that it was easier to make the lantern with a mahogany frame fixed in it and merely to add the glass slide to it. Thus the slide carrier was born. Once it became a normal fixture in the lantern, a standard size had to be adopted for the slides. In Britain 3¼ by 3¼ inches (83 by 83 mm) was chosen as the standard glass size, although continental and American slides were made to a standard size of 4 by 3¼ inches (102 by 83 mm). The image on the glass slide had to be protected from being scraped in its passage through the slide carrier so a second piece of glass, known as the cover glass, was placed over the image glass and bound in place with

gummed tape. These seemingly delicate objects of glass and paper are surprisingly resistant to deterioration by time, although if breakage allows air to penetrate the sealed unit then the picture can be ruined.

Various methods of producing slides were introduced as technical innovations were made. The cheapest method was by lithography, where transfers were printed and placed on the glass. Millions of slides were produced in this manner and they can still often be found in their original boxes in antique shops.

The firm of Butcher and Son of London was one of the biggest manufacturers of this type of slide, their products being sold under the trade name of Primus. The slides were usually sold in boxes of eight and subjects ranged from comic nursery stories to more serious topical subjects like the Boer War. Some boxes of slides seem in retrospect amusing as examples of the ultimate in precis, such as when great books are condensed into eight paragraphs. The Primus slides sold initially for about two shillings per box but by the end of the First World War the price of each set had more than doubled. Other companies produced similar slides and, in an effort to capture the market,

Some boxes of lithographed slides which were intended for children. These brilliantly coloured slides were cheap compared to their hand-painted counterparts and consequently they sold extensively.

made sets of larger numbers (usually twelve) or even pirated each other's stories.

The other main method of mass-producing slides was by photography, and one could buy lantern plates which were photosensitised so that any photograph could be produced as a slide rather than a print. The glass photographic plate was exposed in the usual manner and then developed in the darkroom to produce a negative. When all the negatives were developed they were usually graded by the photographer according to their densities. Still in the darkroom, one plate was taken from a box of commercially produced photo-sensitised lantern plates, placed below a negative and fitted into a wooden frame to prevent movement. A light was then shone through the clear areas of the negative, thus exposing the lantern plate below it and producing a positive in the form of a glass transparency. When the

OPPOSITE AND FOLLOWING PAGE: *A set of slides manufactured by the London firm of York and Son, illustrating the poem, 'Curfew Must Not Ring Tonight'. The narrative verse describes how Bessie saves her lover, doomed to be executed at the toll of the curfew bell, by hanging on to the clapper. Reunited with her lover, Bessie delivers his pardon.*
BELOW LEFT: *The climax of a set of slides entitled 'Blossom and Fruit'. It depicts the burning down of the Leopard public house. Note the use of the painted backdrop and simple stage props.*
BELOW RIGHT: *A scene from a set of slides entitled 'Alice Leighton'. Here the slides have been shot 'on location'. There were thirty-six slides in the complete set.*

slide was fixed and dried a cover glass was fitted and it was bound and ready for use. Almost any amateur photographer could, therefore, develop his own lantern plates and produce his own slides.

The most popular slides in Victorian Britain were those termed 'life model slides'; these mainly depicted popular sentimental stories of the day and were made by various companies. The slides were produced by arranging costumed actors in front of a painted backdrop. Various stage props would be placed in front of the actors and the scene from the story would be photographed. Further scenes would be created in a similar manner until the story was completed.

The slides would often be hand-tinted with colour and sold in boxes, along with a text to be used as a reading. Many of the life model sets were moral tales and were sold to religious societies. In particular, the temperance movement used the slides extensively to convert a mass audience. Many of the melodramatic slide stories were told in verse, the works of George R. Sims proving particularly popular. Slides depicting the words of hymns were produced so that religious societies could conduct whole services by means of the magic lantern. Popular songs of the day were also produced in slide form and these 'pictorial concerts' were always enthusiastically received.

Before the invention of the cinema, the rudiments of moving pictures were explored by magic lantern slides. Although glass is inflexible, the appearance of movement could be effectively achieved for short periods.

The slip slide comprised a fixed piece of glass over which a movable strip of glass could be passed. This second strip was often a black mask which could slowly cover or uncover parts of the fixed picture. Variations on this theme could, by means of small brass levers, move the second piece of glass in a vertical rather than a horizontal plane. Circular movement on slides was more complex but was often achieved by means of a rack and pinion movement, the most obvious application of this principle being a slide of a windmill with revolving sails. More

ABOVE: *A lantern slide depicting a temperance meeting in 1892. Temperance societies were amongst the biggest users of magic lanterns in the late nineteenth and early twentieth century.* BELOW: *Part of a magic lantern programme from the Central Hall, Newcastle upon Tyne.*

Tuesday, March 7th, 1911.

GRAND LIME-LIGHT LECTURE

ENTITLED:

"A Trip on a Trawler on the Northern Seas."

: : Will be delivered by : :

Mr. TOM WING, Ex-M.P. for Grimsby.

Illustrated by numerous specially prepared slides.

Admission 3d.; Reserved Seats (front of Gallery), 6d.

———— Commence 7·30. ————

LEFT: *Portrait of G. R. Sims in 1895. Sims did not see himself as a writer of material for lantern slides but his narrative sentimental verse was used extensively by nearly all 'life model' slide companies.*
RIGHT: *A poetry book by G. R. Sims. Much of the verse in the book has been used for lantern slide stories; the best known verse today is 'In the Workhouse on Christmas Day'.*

ingenious examples could make a sleeping gentleman swallow numerous rats or, with a more educational theme, show the motion of the planets in their orbits or a simulated eclipse of the sun.

Some elaborate and well engineered slides could produce startling effects and often the slides themselves bore impressive names. A double circular movement of geometric patterns produced a moving kaleidoscope effect on the screen. Victorian children nicknamed these slides 'artificial fireworks' but their true name is Chromotrope.

More dramatic geometric designs could be produced by a slide called the Cycloidotrope. This etched complex cycloid curves into smoked glass, throwing the image on to the screen as the curve was being drawn.

Many special effects, however, were produced not by the slides but by the use of more than one lantern; in this way

images could be superimposed on the screen. Specialist apparatus was produced to create these effects. Most notable was the *biunial* (two in one) lantern. The biunial enabled two slides to be precisely aligned, thus allowing one slide to dissolve into another, so creating specially designed dramatic effects (for example, changing winter into spring on a picture). Other effects produced by the biunial could, for example, blend small images over the top of larger pictures to create 'thought pictures' or illusions such as the appearance of angels. The biunial was expensive because essentially it was two lanterns fused into one, although it was argued that in the single lantern all the requirements were present to produce the biunial's effects, provided a way of superimposing two slides in the same carrier could be achieved.

Various complex devices were invented to produce the dissolve effect with

A matching pair of effect slides, entitled 'The Angel of Mercy'. The different images are designed to fit a biunial lantern and produce a startling change as they dissolve into each other.

The Cycloidotrope was a complex instrument which etched patterns into smoked glass. Such a slide cost thirty shillings in about 1900.

The Choreutoscope exploited the phenomenon of persistence of vision in the human eye. The use of the shutter gave the illusion of the skeleton dancing.

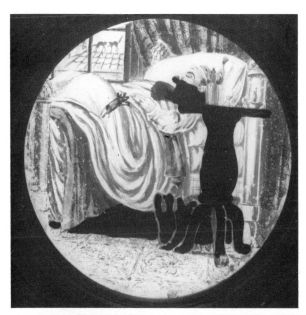

LEFT: *A close-up of a rack and pinion slide. One static and two movable pieces of glass create the illusion of the sleeping gentleman swallowing the rat.*

BELOW: *A superb hand-painted rack and pinion slide of a windmill. Turning the handle causes the sails to revolve.*

Wind-mill.

one lantern. The most successful and popular contender was the Beards eclipse carrier. Despite its claim to produce the dissolve effect with one lantern, the Beards carrier fell well short of this, but it did allow slides to be placed in the lantern from one side only, rather than by leaning over the lantern to place the slide in the opposite slot as in a traditional carrier.

Of all the millions of slides produced by the lantern companies, one is worthy of special note as its mechanics proved to contain the essence of the motion picture. The slide was called the Choreutoscope

and consisted of slightly different images (usually a dancing skeleton) which were cranked by a handle past a small window in the lantern. Each image was seen for a fraction of a second, then a shutter came down while the next image was being moved into place. The rapidity of the cranking gave the impression not of separate images but of continuous movement.

Within a few years a flexible transparent material was to replace the glass and effects of almost infinite duration could be achieved. The motion picture had been born.

The biunial or double lantern allowed complex special effects to be projected on the screen.

The large Ross magic lantern and epidiascope. A complicated system of movable mirrors and lenses inside the body directed the light to the section of the instrument in use at the time.

THE END OF AN ERA

The magic lantern did not cease to be used as soon as the movies began to flourish: for many years the two co-existed. The early cinematographs still required elements from the magic lanterns (such as light source and lens systems). The old-style carrier and the glass slide were, however, replaced by the film mechanism. Many early hand-cranked movie projectors were merely add-on attachments to the front of the magic lantern. The Biokam, for example, was fitted on to the lens flange of an ordinary magic lantern after the objective lens system had been removed. In this way the lantern doubled as a cinematograph, although the device was still in its infancy. The worst defects in this design were that the holes to crank the film through the projector were down the centre of the film and the metal bar which produced this movement tended to scratch the film. No one had yet thought of adding a take-up spool, so the film either poured on to the floor or tumbled into a velvet bag.

Magic lanterns still continued to be made but these were for education rather than entertainment purposes. The elaborate ornamentation and craftsmanship of the Victorian models was gone; the electric filament lamp replaced the limelight and the metal bodies were factory produced, usually finished in black crinkle paint. Typical of these later lanterns was the large Ross lantern/epidiascope, which was purely a functional instrument, having little aesthetic appeal. Nevertheless, the Ross lantern became popular in many schools and colleges.

More and more, film began to replace glass for projection purposes. The great days of the magic lantern were over. The millions of glass slides produced in the Victorian era were mostly consigned to the rubbish tip as they fell out of fashion after the First World War. A few slides were maintained in the cinema palaces as trailers for forthcoming events or advertisements but by the 1950s these too had disappeared.

The Biokam hand-cranked projector. Like many early cinematographs, this was merely an attachment for the front of a magic lantern.

FURTHER READING

Many books about magic lanterns were published in Victorian times, but these are now out of print and difficult to obtain. More recent publications include:-

Bartholomew, C. *Mechanical Toys.* Hamlyn, 1979.
Freeman-Sayer, P. and C. *Victorian Kinetic Toys and How to Make Them.* Evans, 1977.
Gordon, Colin. *By Gaslight in Winter.* Elm Tree Books, 1980.
Household, G. A., and Smith, L. M. H. *To Catch a Sunbeam.* Michael Joseph, 1979.
Mackay, James. *Nursery Antiques.* Ward Lock, 1976.
Smith, L. M. H. *The Road to the Isles.* Macdonald, 1983.

MAGIC LANTERN SHOWS
AND SOCIETIES

Some lanternists put on travelling magic lantern shows in their locality. Details may be obtained by telephone in the following areas:-
Avon: Century Image Magic Lantern Theatre. Telephone: Bristol (0272) 655952.
Northumberland: A Thrill in the Dark. Telephone: Alnwick (0665) 711637.
Sussex: The World of the Magic Lantern. Telephone: Rustington (0903) 782883.
Warwickshire: Victorian Lantern Show. Telephone: Stratford-upon-Avon (0789) 731229.

OTHER USEFUL ADDRESSES
Magic Lantern Society of Great Britain, c/o The Secretary, Prospect, High Street, Nutley, East Sussex.
Magic Lantern Society of United States and Canada, c/o The Secretary, 98 Charles River Road, Watertown, Massachusetts 02172, USA.

PLACES TO VISIT

Intending visitors are advised to find out the times of opening before making a special journey.

Magic Lantern Theatre and Museum, Church Street, Monmouth, Gwent. Telephone: Monmouth (0600) 3146.

Museum of Science and Engineering, Blandford House, West Blandford Street, Newcastle upon Tyne. Telephone: Newcastle upon Tyne (0632) 326789. (A large collection of magic lanterns, not always on display.)

National Museum of Photography, Film and Television, Prince's View, Bradford, West Yorkshire. Telephone: Bradford (0274) 727488. (Magic lantern items not always on display.)

Science Museum, Exhibition Road, South Kensington, London SW7 2DD. Telephone: 01-589 3456.